Monkey Business

Critical flaws in the theory of evolution...

by
Dan Vis

ISBN: 978-1-958155-11-0

Published by FAST Missions
111 2nd Street
Kathryn, ND 58049

Additional copies of this book are available by
visiting us at WWW.FAST.ST

Dedication

This book is dedicated to my secular university instructors, who--while they were all strict rationalists--were also brilliant scholars and taught me to analyze science objectively, and to think, in a way that actually renewed my faith.

Table of Contents

Chapter 1:   The Question of Origins 1

Chapter 2:   Battle Lines 6

Chapter 3:   Science Falsely So Called 13

Chapter 4:   Five Problems with Evolution 21

Chapter 5:   An Unlikely History 29

Chapter 6:   A Faith that Endures 35

Contents

Chapter 1 ....

Chapter 2 6

Chapter 3 12

Chapter 4 17

Chapter 5 19

Chapter 6 25

Monkey Business
Preface

There's a problem trickling through much of Christianity. You can see it in the decaying spirituality of many congregations. Our declining membership. Our loss of young people. If we are honest about it, it's hard to deny our foundations are slowly crumbling.

And the cause? I believe it's for a very simple reason: faith is being undermined, day in, day out, by the proponents of the theory of evolution. It's everywhere. And touted with a confidence that seems invincible.

And we're not really talking about it. Even those who believe the biblical account of creation, rarely acknowledge the conflict. We keep our position out of sight--almost as if we were embarrassed by what the Bible teaches.

But it really shouldn't be that way. The reality is, that the theory of evolution is deeply flawed--in ways that make it wholly untenable. Rather than being intimidated by the supposed overwhelming scientific evidence in its favor, we should be bold and confident, exposing its massive, glaring problems.

In my post-graduate study, I did extensive research into the history and philosophy of science--because I wanted to resolve these questions for myself, once and for all. I discovered that science is not as objective as most people think. Rather, it is frequently distorted and twisted by strong cultural influences.

It's the thesis of this book, that if you want to evaluate evolution objectively, there's only one logical conclusion: it completely flunks the science test!

In the pages ahead, we'll highlight some of those flaws, and give you the information you need to help rebuild the crumbling foundations of our modern church. May it serve to strengthen your faith in the eternal Word, and cause the science falsely so-called of our day to dissipate into the gossamer threads it really is.

The Question of Origins
Chapter 1

In this chapter, I want to talk about a topic we don't often discuss much in Christian circles. And that's the relationship between faith and science. And in particular, the question of origins.

Faith and science have a pretty contentious relationship sometimes, don't they? The Bible is explicit that life in this world was created by the Word of God "ex nihilo" (out of nothing). That it occurred in the recent past (roughly 6000 years ago or so). And that the various animal species we see in the world today were all created, fully developed.

You know the stories: Adam and Eve, the Garden, the Serpent, Cain and Abel, and all the rest.

The scientific community by and large argues something different. That man came about as a result of chance, or accident—through a gradual process over a long period of time. A process called "evolution". A theory first proposed by Charles Darwin, and now taught by scientists all over the world.

And these two explanations of our origins simply aren't compatible. But why? Doesn't nature and Scripture come from the same divine mind? Shouldn't the study of both lead to the same conclusions?

And how should believers respond to these kinds of questions? Does the Bible offer principles to help us navigate such conflicts?

Not An Expert!

Now, before getting too far into this, I want to make it clear I'm not going to try and pretend to be some great scientist. Or a great theologian for that matter! But that's a good thing—as you don't have to worry about this class getting excessively technical.

I have however spent a good bit of time exploring this issue. In both my undergraduate and graduate degrees, I did strong minors in the history and philosophy of science and then did a year or two of post masters research focusing full-time on science/faith issues specifically—at one of the top-ranked (secular) universities in this field.

I was a relatively new Christian back then, and had lots of questions, like a lot of people do. And I really wanted to get to the bottom of those questions for myself.

I wasn't raised in a Christian home, and still remember, while feeling drawn to the moral principles of Christianity, also having doubts as to whether or not it was all true. I was always a thinker, I suppose. A bit of a skeptic, perhaps. And I wanted to be sure. I needed some kind of objective evidence the Bible was true—before I could fully commit myself to that book.

Ultimately, I found that evidence in Bible prophecy. Evidence that was convincing, and compelling. And it led me to faith.

But I still wasn't sure what to do with the Bible record of creation. It was something I personally needed to study for myself. I wanted to understand the issues. To think through the evidence. Which is why I spent so much of my educational program exploring this exact topic.

So, while I'm not an expert, I have definitely given it all some thought!

What's at Stake

I've also come to the conclusion these issues are important. Ignoring the question of origins is not really an option. We can't simply pretend this conflict doesn't exist.

In evangelism, in fact, it's one of our biggest hurdles. This is especially true when trying to reach students at secular college campuses—where I worked for a number of years, and pretty much cut my teeth in ministry.

I still remember a young college student named Jaimbo. He was a smart, entirely impressive young man from mainland China, working on an MBA program at the university I attended. I had met him while going door to door, and he had been open to Bible study—mostly out of basic curiosity about Christianity. So I began to take him through a series of studies on prophecy—and he gradually became pretty convicted. He was astonished at the Bible's ability to predict the future, and with such incredible accuracy.

But he had been taught the theory of evolution all through his childhood, and he had a hard time shaking free from it. He kept asking: "But what about evolution?" At the time I wasn't really equipped to answer that question, and sadly, he ended up not taking a stand for the Bible.

I know now there are actually really simple answers to that question he kept asking. No Christian has to settle for a life of nagging doubt. They can have rock-solid confidence in both the Bible and science—without having to reject either. And they can be armed to give a reasonable answer for their faith. I'll explain how over the next few days.

I just wish I had known this information back then...

The Question of Origins
Worksheet

The Bible teaches explicitly that the_____ was created by the

_____ _____ _____

The scientific community teaches man came about as a result of
_____over a _____ _____ _____ _____

How are these two views incompatible?

Why is it important to NOT ignore the question of origins?
What do we risk if we do?

Do you have to be an expert in science to be able to defend faith
in the Bible? Why or why not?

Additional Notes:

Battle Lines
Chapter 2

The battle lines really are drawn, aren't they? That line between science and faith on the question of origins. And both sides are pretty well dug in, and unwilling to budge.

Before getting into the actual nuts and bolts of the scientific evidence, let's take just a moment to review some of the things each side has going for it:

The Science Side

The scientific, secular, academic, worldview actually has a number of significant strengths on its side. The scientific method seems objective. It's rooted in raw empirical data, not opinions or fanciful speculation. The professors I studied under were brilliant and incredibly knowledgeable. Yet at the same time, manifested an intellectual modesty and reserve I found extremely appealing.

Scientists have been able to generate an incredible amount of information about the natural world. And often have access to amazing labs, where they are able to do all sorts of incredible real-world experiments. And there's a powerful peer review process, by which any scientist's work can be challenged and debunked by any other scientist anywhere in the world.

Also on their side, is our modern-day explosion of technology. The result of our ever-expanding scientific knowledge has been a world filled with technological devices that work in a way that seems almost like magic. In fact, much of science has become so sophisticated and complex—it's almost impossible for the average person to discuss, much less refute the latest developments.

There is no question about it: science comes across in a way that is rather impressive. It's no wonder Christians often feel a bit intimidated by the powerful and generally united claims of the scientific community.

The Bible Side

However, let me take a moment to point out, that the Bible has at least a few things going for it as well.

Billions of copies of the Bible have been printed, and in far more languages, and more places than any other book. Without a doubt, no scientific work comes close to this in comparison. In fact, no other book of any kind. Ever.

And this popularity of the book is largely because readers have confidence in it. In one recent survey of US adults, 45% claimed they believed the Bible's account of creation, while only 10% professed to believe the scientific theory of evolution. (The rest were unsure or undecided). Now, just because a lot of people believe something, doesn't make it true—but this survey certainly suggests something about the credibility many people ascribe to this book.

In fact, people not only believe this book—but countless numbers have sacrificed for the book. Often leaving their homes, to travel to remote and potentially dangerous parts of the world, to share its teachings. These missionaries often have faced hardship, severe persecution, and even death. What science book can you think of that engenders that kind of loyalty?

And why? Presumably, it's because the Bible answers the most important questions of life, and in a really powerful way. This is amazing in its own right: primitive shepherds and fishermen from the distant past seem able to give more compelling answers to man's search for meaning than the most highly educated scientists of today. Why is that?

There is also the evidence of changed lives. Millions through the centuries have experienced dramatic transformations as a result of discovering and reading this book. I am one of them—though there are countless other similar testimonies. Clearly, there's something special about this book. Something as equally inexplicable as technology.

And then there's the evidence of Bible prophecy. The same evidence I shared with my Bible student from China all those years ago. I can't go into that topic here—but the Bible is full of amazingly detailed prophecies about the future that have been fulfilled in remarkably accurate ways. In ways that are inexplicably improbable...

Prophecies about the rise and fall of one world empire after another. Countless details about the life of Christ, all perfectly fulfilled in the life of one man. And numerous passages describing the last days—all uniquely describing things happening right before our eyes right now. The probability of all these various prophecies, written by so many different writers, over so many centuries, all being perfectly fulfilled, by chance—is, quite simply, inconceivable.

So in other words, the Bible has a few things going for it as well!

Who Will Ye Choose?
Comparing these two, it's not completely unreasonable to put

faith in a book like the Bible. When confronted with this kind of conflict, it's not entirely unreasonable to simply accept the Bible by faith. After all:

> *Hebrews 11:3*
> *Through faith we understand that the worlds were framed by the word of God, so that things which are seen were not made of things which do appear.*

Never forget this point: when you're convinced the Bible is true, it's always appropriate to respond in faith to things you don't understand.

But the problem is that all too often we stop there. Or to put it differently, the problem is what many Christians have done with the evolution side of the issue. Their logic goes something like this:

1) I believe the Bible is true.
2) I can't refute evolution.
3) So I think I'll just ignore it.

Yes, choosing to stand on the Bible is commendable. But simply pretending a problem doesn't exist—not so much. How much better to get real answers to our questions? To eliminate that nagging doubt? To resolve our inner intellectual tension? And to shore up those weak spots in our apologetics?

Sometimes, we can win the battle for souls by simply pressing the evidence on the Bible side: changed lives, prophecy, internal consistency, personal testimonies, etc. And this can be convincing even without addressing the claims of Evolution.

But there's also a place for demolishing some of the opposing arguments. For showing the flaws in the theory of evolution. For showing how the scientific evidence actually fails to support it.

The theory of evolution has crippled the faith of millions. And there's a time to cast down "every high thing that exalteth itself against the knowledge of God" (II Corinthians 10:5).

I personally believe this is actually a special part of the work God has given to His end-time people. The three angels messages of Revelation 14 to go "to every nation, and kindred, and tongue, and people" include a call to "worship him that made heaven, and earth, and the sea, and the fountains of waters" (Revelation 14:6-7).

How can we really give that message, if we're unable to answer people's honest questions about our origins?

Fortunately, it's not all that hard. More on that in the next chapter!

Battle Lines
Worksheet

Summarize in your own words the main advantages of the scientific community, that help to give it significant credibility:

Do you ever feel intimidated by the pronouncements of scientists?

Summarize in your own words the main advantages of the Bible, that help to give it credibility among believers:

How strong is your personal confidence in the Bible?

Do you see it as reasonable or unreasonable to believe the Bible over science? Why or why not?

How important is it to be able to give an answer for our faith in Creation? Why is it not enough to simply ignore how widespread evolution really is?

Why do you think God includes a call to worship the Creator in the three angel's messages? In the final warning to planet earth?

Additional Notes:

Science Falsely So Called
Chapter 3

Let's start our study of the theory of evolution by taking a few moments to explore the scientific method. Notice this statement from the book Education, that opens the chapter entitled Science and the Bible:

> *Since the book of nature and the book of revelation bear the impress of the same master mind, they cannot but speak in harmony. By different methods, and in different languages, they witness to the same great truths. Science is ever discovering new wonders; but she brings from her research nothing that, rightly understood, conflicts with divine revelation. The book of nature and the written word shed light upon each other. Education, p 128*

There's a lot in here—let's take a few moments to unpack some key points:

First, the Bible and nature both have the same source. The same Author, the same Creator. And as a result, it makes sense that they would reveal the same character. That they would speak in harmony with each other. While they may use different methods (and this is an important point)—they both bear witness to the same underlying truths.

Second, science—when done properly—will never bring forth anything that conflicts with divine revelation. It's also worth noting that our theology needs to be done correctly too. That is, if we come up with faulty interpretations of the Bible, those views might well conflict with legitimate science. But in those cases, the problem there would lie with religion.

Or to put it differently, science and theology both "shed light upon each other". That is, good theology can help us to interpret the natural world, and good science can help us understand Scripture. Surprising, perhaps, but that's what the quote says. Powerful thought.

Now if it's true that science and faith should always reinforce each other, why are so many scientists convinced we came about through the process of evolution, and not creation?

I mean the Bible is pretty clear on this point, right? There's not a whole lot of theological ambiguity. God created man. Adam and Eve. He put them in the Garden of Eden. There's the fall, and then later the flood. And all the rest. The biblical record is pretty straightforward. Pretty inescapable.

Yet science rejects that and supports instead Darwin's theory of evolution. And it's not the only area of conflict. I won't be able to go into all of them here in this book, for the sake of time, but there's a similar dynamic at play in each of these conflicts—whether it's biology, psychology, geology, or whatever.

Science does seem conflict with the Bible at times. Why?

The Scientific Method

Remember that quote that talked about science and theology each having their own methods and languages? And how there is no conflict when those right methods are used?

Let's take a quick look at the legitimate scientific method.

It's both simple and powerful—and you can find this basic process taught in any foundational science textbook:

First, propose a hypothesis about the natural world. Second, construct an experiment to test that hypothesis. Third, if the experiment fails, change your hypothesis. Basically, it's your hypothesis that fails. And fourth, if the experiment succeeds, hang on to your hypothesis, tentatively.

I say tentatively because there could always be another hypothesis that equally explains the result of your experiment. One that predicts even more results, more accurately. In other words, an experiment does not actually prove a hypothesis is right, it can only prove a theory is wrong.

Finally, if a hypothesis still holds true, after many, repeated, verified experiments, then that hypothesis generally becomes accepted as a theory. Meaning it is generally believed to be true. But even then, it is still not guaranteed and subject to change.

Take the ancient great philosopher Aristotle for example. He developed a theory that everything we saw consisted of four fundamental elements: earth, water, air, and fire. And that each of these elements gravitated toward its natural place. And this theory was validated by all kinds of experiments.

Smoke from a fire went up. Rain fell down. Blow bubbles underwater, and they float up. Drop a rock in water, and it will fall. For more than 2000 years, this theory was widely accepted—because it matched the results of every test you threw at it.

In the 1600's, a man by the name of Isaac Newton developed a new theory: Gravity. He created a simple formula that not only explained all the same phenomena, but did it more elegantly, and more accurately. And Aristotle's theory was quickly discarded. The new theory simply did a better job of matching the observable data.

For the next 300 years or so, Newton's theory took everything thrown against it—quite well. Until Einstein came along, proposing an entirely new way to think about gravity involving the warping of time and space. It not only accounted for Aristotle's observations, and Newton's too, but it explained certain things none of the preceding theories could account for. And so Newton's theory had to go. Or at least, it had to be revised.

The same story can be found in pretty much every scientific discipline. Each field is littered with a trail of discarded theories. Why? Because experiments simply can't really prove a theory is right (at least not absolutely). They can only prove a theory is wrong. They may suggest one hypothesis is better than another, but they can never say there's not a better one still. Right?

That's the scientific method. And the secret to its power is the fact that it has a self-correcting mechanism built in. Before a theory can be widely accepted, it must be tested and verified repeatedly, by multiple people, and multiple times, in multiple locations. And it just takes one failed experiment to disprove a widely held belief. If a theory doesn't match observational data—the theory has to go.

And it has definitely produced incredible results. I mean think of the technological advances of our day: it's all built on wildly advanced scientific theories that have been repeatedly tested and tried. Whether you are talking about chemistry, physics, medicine, electronics, or whatever.

And these theories pretty much always work. They have to work, in fact, because our modern devices rely on these theories. They are used over and over again, by thousands all around the world. Users count on this technology to do what it is supposed to do—consistently. The physical theories underlying these devices are being tested billions of times a day.

Mix these two chemicals together and scientists can

predict exactly what will happen. Roll a ball down an inclined plain and scientists can predict exactly how fast it will go and how long it will take to get to the bottom. Connect two wires and you can predict exactly how much current will be flowing through the circuit. Science enables you to predict exactly what will happen—and then if those results are not achieved, the theory has to go. Seriously, one failed experiment is enough to demand an update or revision.

Now here's the interesting thing: there are no real contradictions with the Bible in any of these disciplines I've been talking about. Wherever the true scientific method is used—it just doesn't create a problem for the believer. Never.

The Theory of Evolution

Now follow along with me on this. Take the theory of Evolution for example, which explains how various life forms (species) came into existence. Does that even remotely appear to follow the scientific method?

I mean, a person can propose a hypothesis about how animals evolve, but can he then go into a lab, design an experiment, and then actually see that evolution happen?

Is it an experiment that is reproducible? Is it an experiment that can be done again and again, by anyone, anywhere—and each time you get the exact same result—a brand new species right there in your lab?

Of course not! It's not the same kind of thing at all.

I had a friend in graduate school—also studying the history of science. And he was a very sharp, very godly young Christian student. At the time, he was doing his whole thesis on this exact point: that in disciplines where there is an apparent conflict between science and religion—like evolution, the age of the earth, geology (whether or not there was a flood), and so on—none of these problem areas actually used the scientific method. They were not really science at all. They do not involve testable, falsifiable, reproducible experiments.

Rather, they all deal with things in the distant past. They are not theories you can prove or disprove in a lab. They simply do not lend themselves to the scientific method.

And if you think about it, it's true. Anytime there is an apparent conflict between science and Scripture—it is never really science. It is always something that just looks like science.

Notice Paul's warning about this exact point:

I Timothy 6:20-21
20 O Timothy, keep that which is committed to thy trust, avoiding profane and vain babblings, and oppositions of science falsely so called: 21 Which some professing have erred concerning the faith. Grace be with thee. Amen.

Did you catch that? What we need to guard against is the opposition of science "falsely so called". The problem is not true science—but fields that merely look like science.

I don't know about you, but I actually find this pretty amazing–that the Bible actually warned us about this explicitly!

In the next chapter, we'll look at five major problems with the theory of evolution. Stay tuned...

Science False So Called
Worksheet

Explain in your own words why science and religion should never contradict each other?

Explain what causes apparent contradictions. Is the problem always with science? Or can it also sometimes be with religion? What do you think of the statement that science and theology both "shed light upon each other"?

Summarize in your own words the principle components that make up the Scientific Method:

Do experiments ever truly prove a theory is correct? If not, what purpose do they actually serve? Give examples if you can:

Does the theory of evolution follow the scientific method? What makes it different from most other scientific theories?

Why does Paul warn us to be careful about "science falsely so called"? How can it lead believers into error?

Additional Notes:

Five Problems with Evolution
Chapter 4

I've actually read Darwin's book, the Origin of Species, from cover to cover. And I can tell you there are major problems with it. Darwin himself admitted his theory had many weaknesses, but his hope was that people would come along later who could add to the theory, and solve some of its most pressing issues.

But the reality is, those problems have instead just gotten worse over time. To where now, today, Darwin's theory is entirely implausible by any reasonable scientific standard. Rather than building support for evolution, the last century and a half have only stacked up the evidence against it.

Let's look at some of those weaknesses. Some of the "big picture" issues that show how desperately problematic the theory of evolution really is. Here are five of them:

1. Evolution has no empirical evidence

Here's how a respected Geneticist from Caltech put it: "Within the period of human history we do not know of a single instance of the transmutation of one species into another one . . . This must be admitted."

What is he saying here? Simply that we've never directly seen evolution happen in the real world. Yes, there is some variation within species, but never changes between or beyond species. Which of course is what the Bible says—that each species would reproduce after its own kind.

We do see clear evidence of micro-evolution. Small changes. Adaptability. But never macro-evolution. That is, we've never seen a dog evolve into a Cat, a Cow, or a Cricket—or anything else for that matter. Dogs only give birth to dogs.

And that's a problem.

As we explored yesterday, science is supposed to be built on objective, empirical evidence. On things you can observe directly. On experimental data. Experiments are required because they help to validate or invalidate a hypothesis.

But in the case of evolution, we don't have a single instance of it EVER being demonstrated in a lab. It doesn't even lend itself to any sort of demonstrable test. It's all distant, unreproducible history.

So evolution is a hypothesis with zero empirical evidence to validate it. It is completely unproven.

2. The fossil record fails to support evolution

Here's what a Biologist from Harvard wrote: "Since we have not the slightest evidence, either among the living or the fossil animals, of intergrading types following the major groups, it is a fair supposition that there never have been any".

Here is his basic point: if you go through the entire fossil record, you always see clearly distinct species. Some extinct species, yes—but no half species. No intermediate or transitional species. No species caught changing from one form to another.

Worse, we go from microorganisms in the pre-Cambrian fossil layers, to a sudden explosion of billions of fossils, of all sorts. All perfectly well developed right from the start.

Darwin recognized this was a major problem in his day, and to his credit acknowledged that. But at that time, only a small number of fossils had been found and were available for study. Darwin assumed, additional fossils would eventually be found, and that some would show these intermediate species—confirming his theory. But it hasn't happened.

And note that the so-called "missing link" is not just between humans and monkeys—it's between every known species of animal. The missing links are everywhere.

Fossils that show evolving, transitional life forms simply do not exist. Which is another major problem for evolution.

3. Natural selection cannot add new features

Here's how an Australian geneticist expressed this problem: "Recombination . . . merely redistributes existing genetic material among different individuals; it makes no change in it."

To explain his point, let's review a few basics. Natural selection, which is the mechanism Darwin postulated for evolution, is the idea that the fittest members of a population will be more likely to pass their traits down to the next generation. And that over time, the accumulation of these slight variations within a population can lead to entirely new species.

Now Darwin never speculated how new features would come into existence. Rather, he just argued that if they did appear, natural selection would ensure the best ones would be passed on.

And to be fair—Darwin had no real concept of genetics. The best they could do with a microscope in his day was see a dark glob we now call the nucleus. And Mendel was just barely beginning his research on inheritance in pea plants. The discovery of DNA was still far in the future.

But of course, now we know exactly how DNA works. When a male and female come together, they simply mix existing genes. Some genes come from the father and some from the mother. But nothing new is created. It's just a unique combination.

It's kind of like shuffling cards. You can draw a new hand every time, but you can only draw cards that are already in the deck. You'll never draw something entirely new.

So the scientist above was exactly right. We now know from genetics that there is no way natural selection could account for new features. Which literally makes evolution, as Darwin taught it, impossible.

4. Mutations can't account for speciation either

Here's how a Biologist from Berkeley put it: "Microevolution by the accumulation of micro-mutations . . . leads to diversification strictly within the species . . . The step from one species to another requires another evolutionary method than that of sheer accumulation of micro-mutations".

Here's the story behind this statement. Basically, because most evolutionists realize Darwin's theory cannot work as written, they have developed a new theory, called Neo-Darwinism. It says new features are introduced through mutations. That somehow mutations can encode entirely new features into an organism's DNA. But this doesn't really work either.

For one thing, mutations are extremely rare. Take fruit flies. Experiments have been done subjecting thousands of fruit flies for thousands of generations to massive amounts of x-rays to see if they could induce change. And guess what? You still end up with fruit flies. Minor variations, sometimes, but nothing new.

Mutations are also extremely small. Imagine changing one letter in a whole encyclopedia. Now imagine the amount of change required to change a dog into a cat. We're talking millions of changes. Whole chromosomes. The whole encyclopedia has to be rewritten. There's no way mutations could ever do that.

And last but not least, mutations, if anything, are harmful. They are almost always degenerative. They cause the organism with the mutation to weaken, not improve. And if there are enough mutations, the organism usually dies.

To put it differently, if natural selection is like drawing a hand of cards from an existing deck—a mutation is like shooting a bullet through one of the cards. It doesn't create a new card, it just destroys an existing one.

In other words, mutations can't really explain complex new features. Which leaves evolution with no mechanism for change.

5. Evolution cannot explain the origin of life

Here's how a biologist from Harvard put it: "One has only to contemplate the magnitude of this task to concede that the spontaneous generation of a living organism is impossible. Yet here we are—as a result, I believe, of spontaneous generation"

This is perhaps the most difficult question of all: Where did the first life forms come from? Even if evolution could somehow produce speciation, how did life itself get started?

In Darwin's book — he simply evaded the question. On the last couple pages of his book, He actually suggested God created the first few life forms, and then left everything we see today to evolve from there. Which actually makes him a creationist, I suppose!

But he had to say this because Louis Pasteur had just recently pretty well debunked the theory of spontaneous generation. That life could spring forth spontaneously from non-life. So Darwin, not wanting to appear ignorant of the latest research, simply side-stepped the whole issue.

Of course, the problem of spontaneous generation today is far more severe than then. Take the simplest life form: a virus. (We all know about those, right?) They are actually astoundingly complex. They are literally swarming with chemical processes. Just the DNA, involves tens of thousands of letters—and that's only one molecule in a virus, out of thousands.

And that's for a strikingly simple life form. A virus, for example, can only survive by living off other, more complex cells. It can't even reproduce without hijacking another cell's equipment, right? What about something more complex?

Human DNA for example, involves billions of letters. The equivalent of a library with thousands of books, each with hundreds of pages, and hundreds of words per page. And every letter has to be just right! And again, that's just for one molecule, out of thousands, probably millions, required for one human to exist.

Realistically, what is the chance of something like this springing into life by random chance? Mathematically—it is statistically impossible.

Today, there is no viable theory for how life came into being. And whatever Darwin may have thought, if you can't explain the first life forms, the whole theory of naturalistic evolution collapses.

Summary

There are other arguments against evolution, and for creation—we've just scratched the surface. But these should be enough to suggest evolution really is rather problematic:

1) There's no empirical evidence. It's never been observed.
2) There are no transitional life forms in the fossil record.
3) Natural selection cannot introduce new features.
4) Mutations cannot introduce new features either.
5) And lastly, evolution cannot explain the origin of life.

Problems like these would kill any other theory. Yet evolution lives on. In the next chapter, we'll explore why...

Five Problems with Evolution
Worksheet

Take a moment to summarize in your own words each of the 5
problems with evolution outlined in this reading:

Evolution has no empirical evidence.

The fossil record fails to support evolution.

Natural selection cannot add new features.

Mutations can't account for speciation either.

Evolution cannot explain the origin of life.

Based on the information in this reading, would you describe evolution as being supported by the evidence that exists today? Why or why not?

Additional Notes:

An Unlikely History
Chapter 5

I closed yesterday's reading with this quote from a Harvard biologist—and it was pretty insightful:
One has only to contemplate the magnitude of this task to concede that the spontaneous generation of a living organism is impossible. Yet here we are— as a result, I believe, of spontaneous generation.

That's pretty revealing, isn't it? He's essentially saying, it's impossible, yet I believe. And actually, every one of the experts I cited in previous chapter's reading is an ardent evolutionist. They are all saying the exact same thing: it's impossible, yet I believe...

Which raises an interesting question: why do so many cling to this particular theory, if the evidence really does fail to support it? And so spectacularly?

Positivism

Here's some information from the history of science that may help to explain why evolution is so widespread. It also helps to debunk the myth of absolute objectivity in science. At least in the case of Darwin!

The Enlightenment in Europe raised a whole generation of philosophers trying to explain the natural world in rational and scientific ways. One, a slightly earlier contemporary of Darwin, was a French philosopher by the name of August Comte.

He became fairly influential in his day, developing a philosophy called positivism. It was one of the earliest attempts to rigorously define true "science". According to Comte, knowledge progresses through three stages:

1) The Theological. Everything is explained by supernatural forces. It's all basically myth, magic, and superstition at this point.
2) The Metaphysical. These are the early attempts to explain a subject. We speculate about abstract forces, but don't really understand how they work. And then finally,
3) The Scientific or Positive. Ultimately we transition into a well-established, mathematical, experimentally verified, non-religious theory.

Once a theory reaches this final stage, he argued (naively), it can be considered true at last and is no longer subject to change.

According to Comte, some fields had already been positivised: physics and astronomy in particular. Chemistry was well underway. He predicted the "scientific" method would next be successfully applied to biology, then psychology, and last of all, sociology. Once we finally mastered that last discipline—we would be able to build a perfect society. A society he believed, incidentally, would be run by scientists! And it would never need to be changed or improved again. Wow.

Not everyone agreed with all his ideas, of course. But the belief science had to be non-theological, and non-metaphysical, did spread far and wide. It was soon widely accepted, that scientific theories which included supernatural or religious elements, were not really science at all. Period.

One thing that made his theory so attractive was how prophetic it seemed to be. Darwin began the positivization process in biology soon after, with the first non-religious explanation of origins. Next came psychology, with the work of Sigmund Freud. His theory of the id, ego, and superego, was built on the supposed "evolution" of the various structures in the brain. And then Karl Marx proposed a novel sociological theory, that all societal interactions could be explained as a clash between economic classes. It seemed as if everything was escaping religion! The world was becoming scientific, and thereby positive, at last.

Of course, none of these theories really worked all that well. As more and more evidence came to light, it became increasingly apparent evolution flunked the science test. Psychoanalysis splintered into a thousand competing psychological theories, with little agreement between them—each arguing it was right, but offering no way to prove it. The social theories of Marx, when put to the test in communist Russia proved a cultural, social, and economic disaster. These so-called "positive" theories all failed miserably.

Biased by Definition

But here's my point: society was looking for scientific theories to explain everything. Theories that by definition were non-theological, and non-religious. So when evolution was first proposed, it was eagerly adopted by many because of its one unique advantage over creationism: it did not require a supernatural Creator.

This is also the reason evolution is still taught today. It's not that the scientific evidence supports it—rather, it is because the only other explanation (creation) is theological, and thus not scientific by definition.

Granted, it sounds reasonable to argue you can't use miracles, magic, and mysticism, to explain natural phenomena. Science should be objective, rational, naturalistic. But there's a problem.

If you rigidly adopt this definition and try to explain our origins—you are forced to exclude all explanations that involve God. Which is kind of problematic, isn't it? What if there really were a Creator? How would you ever get to that truth? Your definition of science would force you to automatically rule out any options that even remotely suggest that possibility.

Worse, if someone were to offer evidence that supports creation, it would be discounted as non-scientific, merely because it hinted at a Creator. Even if your evidence were sound, your conclusions would be rejected instantly! In other words, there is not even a way to correct a theory like evolution, regardless of how flawed it might be. You never even get to the question of evidence.

Evolution is taught as the only credible theory of origins primarily because it is the only non-theological explanation available. Despite the fact there is 1) no empirical evidence for it, 2) the fossil record fails to support it, 3) natural selection cannot explain new features, 4) mutations can't either, and 5) it provides no explanation for the spontaneous generation of life. And it will continue to be taught, as long as the current definition of science holds.

All the supposed advantages of science: objectivity, fairness, open debate, and all the rest—aren't really present in the evolution/creation debate. It's not even close to a level playing field. Modern man's flawed approach to science has completely stacked the deck against creationism. The scientific community has essentially zipped itself into a straight jacket.

Which is why I'm not particularly intimidated—even as a non-scientist—in suggesting the entire scientific establishment is wrong on the question of origins. And neither should you...

An Unlikely History
Worksheet

Who was August Comte? What was his philosophy of "positivism" all about?

What did he predict science would eventually do?

How did his theory bias scientists against theories like creationism? In what ways were his theories helpful? In what ways were they not?

How does the story of positivism help us understand why evolution is still taught today even though it is not supported by any real evidence, and has many serious flaws.

Additional Notes:

A Faith that Endures
Chapter 6

It's been quite a journey, hasn't it? We've covered a good bit of ground in our exploration of this debate on origins. And by way of quick summary, let's review some of the key principles we've tried to establish so far:

1) While the scientific community can certainly be impressive, the Bible has its own unique strengths. And there's nothing unreasonable about simply believing by faith what the Bible says.
2) Evolution doesn't actually use the scientific method. It's not something that can be tested, experimented on, or validated in any kind of lab. Rather it simply proposes an unverifiable hypothesis about the distant past.
3) Evolution has major problems and flaws that even its founder recognized, and which have only gotten worse, not better, with the passage of time. If anything, the evidence points even more to a Creator today. And...
4) Evolution is taught everywhere primarily because society historically decided to limit science to those theories that exclude God by definition. The problem has never really been the evidence, but rather a flawed definition.

We've hardly exhausted the topic, of course. There are plenty of good scientists pointing out the improbability, even impossibility, of Darwin's theory. Many are Christian, but not all. In fact, a number of significant atheists have had the intellectual courage to speak out and say things that just don't add up!

If you want to study this subject further, there are plenty of great resources out there. A wealth of information on many different facets of the science/faith debate. I encourage you to inform yourself.

In my experience, it doesn't really matter what scientific theory someone holds up as contradicting the Bible—you'll find the same sorts of problems every time. During my years of graduate study in the history of science, I explored pretty much every scientific discipline that seemed to oppose Christianity. Quite a few rabbit holes, believe me! And yet I never found one instance of true science contradicting the correct interpretation of the Bible. It's just not there.

Still, creationists will remain a minority for the foreseeable future. And will continue to face ridicule and scorn for professing faith in a dusty old book. But we can minimize some of that scorn by being able to give an intelligent answer for our faith. By presenting evidence that shows such faith is reasonable. It may not change anyone's mind or the overall situation in our world, but we can try.

And we need to...

An Endtime Message

The Bible warns that this would be an issue just before Christ returns. Jesus hinted at it when he asked the question: "when the Son of man cometh, shall he find faith on the earth?" (Luke 18:8). True faith in God's Word is clearly going to be a rare thing. Which means we need to advocate for it.

Peter was even more explicit about the problems we would face when he prophesied there would "come in the last days scoffers" saying "where is the promise of his coming?" (II Peter 3:3-6). He continues by pointing out the key intellectual planks that lay at the foundation of that skepticism:

1) "All things continue as they were from the beginning". That is, the earth has continued for long ages with no real change. Scientists call this geological uniformitarianism.
2) "They willingly are ignorant of, that by the word of God the heavens were of old". Similarly, they refuse to believe in creation, ex nihilo—out of nothing. And note they do this willingly—intentionally rejecting evidence to the contrary. And...
3) They deny "the world that then was, being overflowed with water, perished". They reject the evidence for a global flood—another flashpoint between science and religion. It's a topic we haven't explored here, but the scientific evidence for a global flood is equally compelling.

When I reflect on this passage, I can't help but chuckle to myself. The more evolutionists push their theory, the more they confirm the Bible's ability to predict the future!

And we do have a moral obligation to share what we know. In Peter's first epistle, he instructs us to "be ready always to give an answer to every man that asketh you a reason of the hope that is in you with meekness and fear" (I Peter 3:15). We don't have to become a scientific expert—but we should be able to show that Christianity is in fact eminently reasonable.

And we need to be able to respond to the challenges posed by science in particular. Paul closed his first epistle to Timothy with a warning to "keep that which is committed to thy trust, avoiding profane and vain babblings, and oppositions of science falsely so called: which some professing have erred concerning the faith." (I Timothy 6:20-21). He knew science was going to be a problem, and that it would be used by the enemy to lead many astray.

Part of our warfare as Christians involves "the pulling down of strong holds; casting down imaginations, and every high thing that exalteth itself against the knowledge of God" (II Corinthians 10:4-5). And evolution is certainly one of those strongholds! We are to "have no fellowship with the unfruitful works of darkness, but rather reprove them" (Ephesians 5:11).

As Christians, it's important to understand that this is not a trivial or unimportant debate. According to the book of Revelation, God's creatorship lies at the heart of all true worship: "Thou art worthy, O Lord, to receive glory and honour and power: for thou hast created all things, and for thy pleasure they are and were created" (Revelation 4:11). Students of Bible prophecy should also recall that worship, and a specific memorial of creation, in particular, are prophesied to play a critical part of earth's final test.

Conversely, we're told the "wrath of God" will be "revealed from heaven" against those who reject truth. Why? "Because that which may be known of God is manifest ... the invisible things of him from the creation of the world are clearly seen, being understood by the things that are made, even his eternal power and Godhead; so that they are without excuse" (Romans 1:18-20). In other words, the evidence is so irrefutable, no scientist will be able to claim they just didn't know.

Fortunately, the Bible predicts God will have a people to give this warning to the world. Under the symbolism of a mighty angel, God's end-time workers will give this proclamation: "Fear God, and give glory to him ... and worship him that made heaven, and earth, and the sea, and the fountains of waters" (Revelation 14:7). In fact, we're told it will go "to every nation, and kindred, and tongue, and people" (Revelation 14:6). The world will be urged to return to worshiping the Creator. All who refuse will be called to account.

It's a solemn and serious topic. Let's take the time to study to prepare ourselves for the crisis just ahead. To stand in our place, and give the message God has called us to give...

A Faith That Endures
Worksheet

How important is it to inform ourselves on subjects like the debate between evolution and creation.

How can giving a reasonable answer for our faith help to minimize ridicule and scorn?

How did the Bible predict this specific topic would be a point of contention in the days just before Christ returns?

What does it suggest our role should be in this conflict?

How is a call to worship the creator central to earth's final warning? Why is this message so vitally important?

Additional Notes:

FAST Missions
Cutting-Edge Tools and Training

Ready to become a Revival Agent? FAST Missions can help! Our comprehensive training curriculum will give you the skills you need to take in God's Word effectively, live it out practically, and pass it on to others consistently.

Eager to start memorizing God's Word? Our powerful keys will transform your ability to hide Scripture in your heart.

Want to explore the secrets of "real life" discipleship? Our next level training zooms in on critical keys to growth, like Bible study, prayer, time management, and more.

Want to become a worker in the cause of Christ? Our most advanced training is designed to give you the exact ministry skills you need to see revival spread.

For more information, please visit us at:
WWW.FASTMISSIONS.COM

Study Guides

Looking for life-changing study guides to use in your small group or Bible study class? These resources have been used by thousands around the world. You could be next!

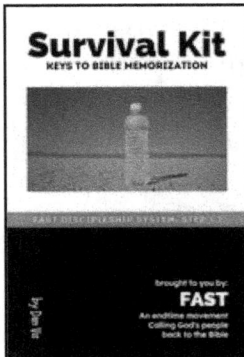

Survival Kit

Want to learn how to memorize Scripture effectively? These study guides will teach you 10 keys to memorization, all drawn straight from the Bible. Our most popular course ever!

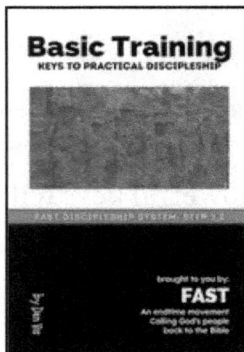

Basic Training

Discover nuts and bolts keys to the core skills of discipleship: prayer, Bible study, time management, and more. Then learn how to share these skills with others. It is the course that launched our ministry!

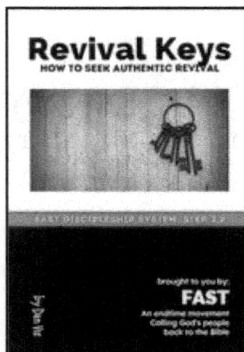

Revival Keys

Now as never before, God's people need revival. And these guides can show you how to spark revival in your family, church, and community. A great revival is coming. Are you ready?

Online Classes

Want to try out some of the resources available at FAST? Here is just a small sampling of courses from among dozens of personal and small group study resources:

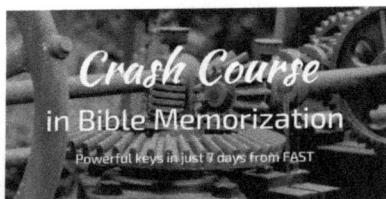

Crash Course
Discover Bible-based keys to effective memorization.
http://fast.st/cc

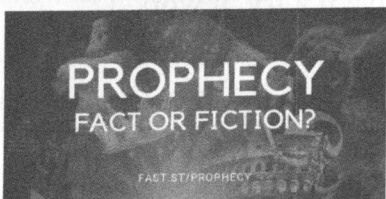

Fact or Fiction
Does the Bible really predict future events? You be the judge.
http://fast.st/prophecy

Monkey Business
Find out how evolution flunks the science test.
http://fast.st/monkey

Revive
Want more of God's Spirit? Learn how to pursue revival.
http://fast.st/revive

The Lost Art
Rediscover New Testament keys to making disciples.
http://fast.st/lostart

Digital Tools

FAST offers a number of powerful "apps for the soul" you can use to grow in your walk with God. And many of these are completely free to anyone with an account. Some of these include:

Review Engine
Our powerful review engine is designed to help ensure effective longterm Bible memorization. Give it a try, it works!

Bible Reading
An innovative Bible reading tool to help you read through the entire Bible, at your own pace, and in any order you want.

Prayer Journal
Use this tool to organize important requests, and we'll remind you to pray for them on the schedule you want.

Time Management
Learn how to be more productive, by keeping track of what you need to do and when. Just log in daily and get stuff done.

For more information about more than twenty tools like these, please visit us at *http://fast.st/tools*.

Books

If the content of this little book stirred your heart, look for these titles by the same author.

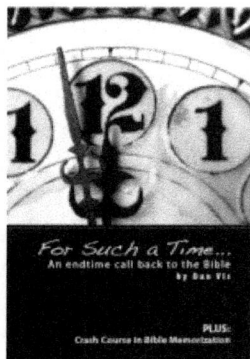

For Such A Time...
A challenging look at the importance of memorization for the last days, including topics such as the Three Angel's messages and the Latter Rain.

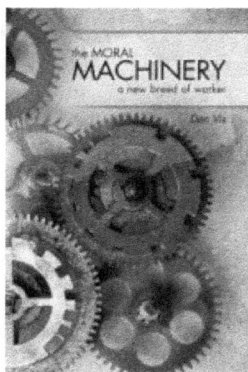

Moral Machinery
Discover how our spiritual, mental, and physical faculties work together using the sanctuary as a blueprint. Astonishing insights that could revolutionize your life!

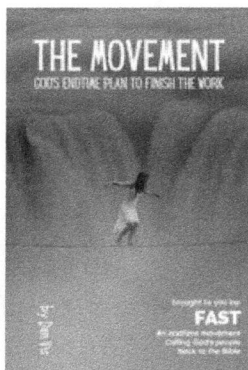

The Movement
Discover God's plan to finish the work through a powerful endtime movement. Gain critical insights into what lies just ahead for the remnant!

www.ingramcontent.com/pod-product-compliance
Lightning Source LLC
Chambersburg PA
CBHW072036060426
42449CB00010BA/2282